WHAT IS... ? See-Through

© 1996 Rigby Education
Published by Rigby Interactive Library,
an imprint of Rigby Education,
division of Reed Elsevier, Inc.
500 Coventry Lane,
Crystal Lake, IL 60014

All rights reserved. No part of this publication may be reproduced or transmitted in any form or by any means, electronic or mechanical, including photocopying, recording, taping, or any information storage and retrieval system, without permission in writing from the publisher.

Cover designed by Herman Adler Design Group
Designed by Heinemann Publishers (Oxford) Ltd
Printed in China

00 99 98 97 96
10 9 8 7 6 5 4 3 2 1

Library of Congress Cataloging-in-Publication Data

Warbrick, Sarah, 1964-
What is see-through?
    p.   cm. -- (What is--?)
Summary: Presents the physical property of opacity in everyday objects.
ISBN 1-57572-051-5 (library)
1. Matter--Properties--Juvenile literature. 2. Opacity (Optics)--Juvenile literature. [1.Matter--Properties. 2. Opacity (Optics)] I. Title. II. Series: Warbrick, Sarah, 1964- What is--?
QC173.36.W42 1996
620.1'1295--dc20   95-41115
                                                                                              CIP
                                                                                               AC

Acknowledgments
The publishers would like to thank the following for the kind loan of equipment and materials used in this book: Spoils, Bishop Stortford.
Toys supplied by Toys Я Us Ltd,
the world's biggest toy megastore.

Special thanks to George, Jodie, Nadia, and
Rose who appear in the photographs

Photographs: Network p15; OSF (Oxford Scientific Films) pp6-7;
Science Photo Library p19; other photographs by Trevor Clifford
Commissioned photography arranged by Hilary Fletcher

There are see-through things all around us.
See-through things let the light through.
See-through things show you what's inside.

This book shows you what is see-through.

These things look different.
What differences can you see?

In one way they are the same.
They are all see-through.

Rose can see through a window.

That's because it is see-through.

The fish tank is see-through.

You can see all the pretty colored fish.

You can see through this glass.

But doesn't Nadia look funny!

What do you think is in
this paper bag?

Apples! You can see them when they are in a clear plastic bag.

These plastic containers are used for storing things.

And you can see what's inside.

You can see through these swimming goggles.

They'll protect your eyes.

Water is see-through.

But it can still make things look very different.

Your hand isn't see-through.

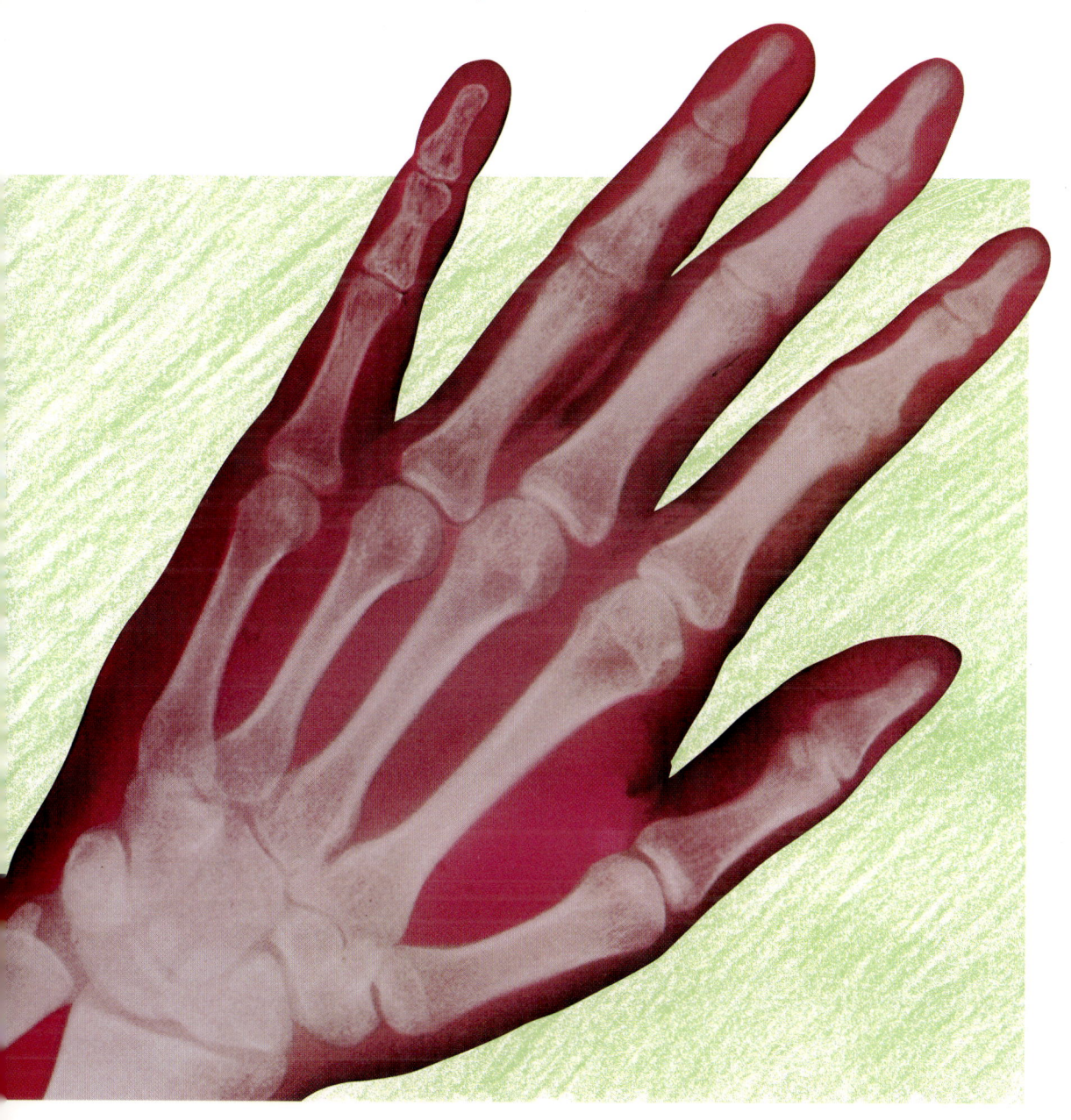

An X-ray can show you what's inside.

What is see-through here?